FLORA OF TROPICAL EAST AFRICA

LYCOPODIACEAE

Bernard Verdcourt

Terrestrial or epiphytic herbs or sometimes growing on rocks, with erect, pendulous, climbing or prostrate, unbranched or dichotomously branched stems. Leaves small, simple, single-veined, spirally arranged or decussate. Sporangia solitary, borne in the axils of sporophylls which either do or do not differ from the foliage leaves or are distinctly different and form strobili or at least strobiliform areas. Sporangia uniform, unilocular, globose or reniform, dehiscing into two valves, producing spores of only one sort. Spores tetrahedral or ± globose with trilete scar. Gametophytes completely mycorrhizal, colourless, tuberous and subterranean or with some chlorophyll and surface-living.

In the world's major Floras only two genera have been recognised until quite recently although specialists have for many years recognised numerous genera and infrageneric groupings. It is now generally agreed that *Lycopodium* must be divided into at least three genera, not really surprising for a group of such ancient lineage.

The Australian genus *Phylloglossum* Kunze has always been kept separate; following Øllgaard (in Kubitzki, Fam. Gen. Vasc. Pl. 1: 31–39 (1990)) *Lycopodium* sensu lato is considered to comprise three genera. I have omitted much synonymy not relevant to our Flora which can be found in F.Z. and other references.

1. Fertile parts forming very dense cylindrical strobili [either solitary or up to 6] at the apices of erect shoots or peduncles from a creeping terrestrial main stem, or solitary at ends of numerous ultimate branchlets of erect stems very densely branched in Christmas tree-like fashion; sporophylls chaffy, totally different from foliage leaves 2

Fertile parts not forming cylindrical strobili but mostly looser strobiliform structures with ± coriaceous or foliaceous sporophylls which are similar to or different from foliage leaves or sporangia borne in axils of scattered sporophylls ± identical with the foliage leaves; one species terrestrial erect but with reduced main stem and others ± pendent epiphytes ... 1. **Huperzia**

2. Strobili (1–)2–6 at ends of erect rather sparsely leafy peduncles; terrestrial .... 2. **Lycopodium**

Strobili solitary at ends of erect peduncles or at ends of numerous ultimate branches of Christmas-tree like erect shoots; terrestrial .... 3. **Lycopodiella**

The following will help with sterile material.

Foliage leaves mostly ± coriaceous; plants ± pendent epiphytes or in one case erect; main basal stems not long .... 1. **Huperzia**

Foliage leaves not coriaceous; plants always terrestrial; main basal stems long-creeping and producing erect shoots 2

Erect shoots much branched in Christmas tree-like fashion or if much simpler then foliage leaves over 1 mm wide . . . . . .   3. **Lycopodiella**
Erect shoots simply branched with leaves under 1 mm wide (at ends of the branchlets the hair-like leaf apices form tufts which often dry orange-brown) . . . .   2. **Lycopodium**

1. HUPERZIA

Bernh. in J. Bot. (Schrad.) 1800 (2): 126 (1801)

Urostachys Herter in Beih. Bot. Centralbl. 39 (II): 249 (1922) (*nom. superfl.*)

Terrestrial, epiphytic or sometimes growing on rocks, erect or drooping; stems equally bifurcate, the basal main stem small and unbranched with roots usually in a single basal tuft or sometimes branches rooting near tips or along prostrate stems. Leaves all similar or different, equal or unequal. Sporophylls identical with foliage leaves or small and/or differently shaped forming distinct strobiliform fertile areas, persistent. Sporangia axillary, reniform with two equal valves. Spores foveolate-fossulate. Gametophytes subterranean or under debris on trunks, colourless, completely mycorrhizal.

A nearly cosmopolitan genus occurring in tropical, temperate and arctic or alpine places with species variously estimated at 200 or 400, and probably about 300 (fide Øllgaard); 8 species occur in East Africa.

Several authors e.g. Tardieu, Fl. Madag. 13 & 13 bis (1971) have placed this genus in a family of its own, Huperziaceae.

1. Plant terrestrial, erect, with sporophylls not differentiated from foliage leaves . . . .   1. *H. saururus*
    Plant epiphytic 2
2. Sporophylls not clearly differentiated from foliage leaves 3
    Sporophylls clearly differentiated from foliage leaves, often forming strobiliform fertile areas 6
3. Leaves linear, 5–6(–10) mm long, 0.2–0.5 mm wide . . . .   2. *H. verticillata*
    Leaves longer or, even if so short, then wider, always > 1.5 mm at base 4
4. Leaves 8–10-ranked, very densely placed, the stems 2–3.5 cm wide (including the leaves); leaves linear-lanceolate (± subulate) up to 2.25 cm long, 1.5–3 mm wide . . . .   5. *H. holstii*
    Leaves not as above, never so long (rarely up to 2.4 cm in *H. gnidioides*) 5
5. Leaves (8–)12–18 mm long, 1.5–2(–3.2) mm wide, densely packed, more linear-lanceolate and sharply acute at apex   4. *H. dacrydioides*
                                           (forms without distinct strobiliform area)
    Leaves not as above, more lanceolate, often more openly placed or bluntly acute at apex . . . .   Juvenile forms of *Huperzia* and forms of *H. ophioglossoides* & *H. gnidioides* without distinct strobiliform areas
6. Leaves spreading at ± right angles, flat and lanceolate (8–20 × 2–3.5 mm) 7
    Leaves erect, appressed or spreading at no more than ± 45° 8

7. Strobiliform areas about 2 mm wide; sporophylls 1.5–2 mm
long, 1.5 mm wide leaving much of sporangium
uncovered and as long as or scarcely exceeding it . . . . .   *7. H. phlegmaria*
Strobiliform areas 3–4 mm wide; sporophylls (2–)3–4 mm
long, 1.5–2 mm wide, covering most of sporangium and
distinctly longer than it (**U** 2) . . . .   *8. H. staudtii*
8. Leaves and often sporophylls rather openly arranged
(2–3-ranked) with the axis clearly visible; leaves
8–12(–15) × 1.3–4 mm; sporophylls ovate-lanceolate,
2–4 × 1–1.5 mm . . . .   *6. H. ophioglossoides*
Leaves and sporophylls tightly arranged, the axis covered 9
9. Leaves bluntly acute or subobtuse, usually more appressed
to the stem, linear-oblong to lanceolate, flat, 9–15(–24)
× 1.5–3(–4) mm; sporophylls ovate-acuminate, 2–5 ×
1.75–2.5 mm . . . .   *3. H. gnidioides*
Leaves more sharply acute, often more spreading, more
narrowly linear-lanceolate, not flat but more inrolled
at edges and appearing ± subulate, 8–12(–18) ×
1.5–2(–3.2) mm; sporophylls very variable 2.5–13(–18)
× 1.8–2 mm . . . .   *4. H. dacrydioides*

1. **Huperzia saururus** (*Lam.*) *Trevisan* in Atti Soc. Ital. Sc. Nat. 17: 248 (1875); Rothm.
in F.R. 54: 60 (1944); Tardieu, Fl. Madag. 13 & 13 bis: 24, fig. 3/5–6 (1971); Pic. Serm.
in B.S.B.B. 53: 185 (1983); Schippers in Fern Gaz. 14: 174 (1993); Faden in U.K.W.F. ed.
2: 38 (1994). Type: Réunion [Ile Bourbon], *Commerson* s.n. (P-LAM, holo.)

Terrestrial or sometimes on rocks; horizontal stems compactly branched, with
shallow roots; aerial stems somewhat fleshy, mostly unbranched or occasionally once
or twice bifurcate, 9–60 cm tall, up to 2 cm wide including the leaves (the stem itself
about 3 mm). Leaves closely imbricated, somewhat spreading or usually adpressed,
narrowly oblong-lanceolate, 8–13 mm long, 2 mm wide, sporophylls not
differentiated from the foliage leaves; sporangia yellow-brown, rounded-reniform,
1.5–2 mm long, flattened, mostly obscured.

UGANDA. Toro District: Ruwenzori, Bujuku Valley, 30 Dec. 1950, *G. Wood* 197!; Kigezi District:
Mt Mgahinga, 4 Dec. 1930, *B.D. Burtt* 2872!; Mbale District: Bugishu, Mt Elgon, 22 Mar. 1951,
G. Wood 104!
KENYA. Elgeyo District: Cherangani Hills, Chemnirot, 16 Aug. 1969, *Mabberley & McCall* 185!;
Aberdares, Kinangop, 17 July 1948, *Hedberg* 1627!; N Nyeri District: Mt Kenya, Naro Moru
track, 8 Sept. 1963, *Verdcourt* 3735!
TANZANIA. Moshi District: Kilimanjaro, N slope of Kibo, 22 Feb. 1933, *Rogers* 788! & above
Peters Hut, 26 June 1948, *Hedberg* 1355!; Rungwe District, N slope of Rungwe Mt, 8 Feb. 1961,
Richards 14297! & 14323!
DISTR. **U** 2, 3; **K** 3, 4; **T** 2, 6*, 7; Cameroon, Congo (Kinshasa), Malawi, Zimbabwe, South Africa,
Comoros, Réunion, Mauritius, Kerguelen Is., Tristan da Cunha and St. Helena; also in S America
HAB. Near streams and pools, sphagnum swamps, boggy moorland, seepage zones in rocks and
exposed grassy crater rims, *Hagenia* forest; 2200–4400 m

SYN. *Lycopodium saururus* Lam., Encycl. Méth. Bot. 3: 653 (1789); Hieron. in P.O.A.C: 90 (1895)
& in V.E. 2: 71, fig. 68 (1908); Sim, Ferns S. Afr. ed. 2: 324, t. 175 (1915); F.D.-O.A.: 88
(1929); Chr. in Dansk Bot. Arkiv 7: 188 (1932); Alston, Ferns W.T.A.: 11 (1959); Tardieu,
Fl. Cameroun 3: 10 (1964); W. Jacobsen, Ferns S. Afr.: 132, fig. 74 (1983); Schelpe &
N.C. Anthony, F.S.A. Pterid.: 5, fig. 2/1 (1986); J.E. Burrows, S. Afr. Ferns: 12, t. 1/2, fig.
2/2, 2a (1990)

* recorded from Uluguru Mts, Lukwangule Plateau by Schippers (1993).

2. **Huperzia verticillata** (*L.f.*) *Trevisan* in Atti Soc. Ital. Sc. Nat. 17: 248 (1875); Rothm. in F.R. 54: 60 (1944); Tardieu, Fl. Madag. 13 & 13 bis: 23, fig. 2/1–4 (1971); Pic. Serm. in B.J.B.B. 53: 185 (1983); Schippers in Fern Gaz. 14: 176 (1993); Faden in U.K.W.F. ed. 2: 38 (1994). Type: Réunion, *Sonnerat* s.n. "per Thouin" (SBT, holo.)

Pendulous epiphyte or occasionally on rocks; rooting stem 4–6 cm long, the aerial stems at first erect but soon pendulous, branching dichotomously several times, the first branching at about 4–8 cm from root; total length 25–50 cm long, 3–6(–10) mm wide near base, 2–3 mm wide at apex. Leaves ± adpressed to spreading, linear, 5–6(–10) mm long, 0.2–0.5 mm wide, with distinct midrib. Sporophylls very similar to foliage leaves. Sporangia can occur all along the stems but mainly in the upper parts, not hidden, 2–2.5 mm wide.

KENYA. Fort Hall District: Tuso, 28 July 1908, *Balbo* 819!; Kiambu District: Uplands, Gatamayu R. [Katamayo], Aug. 1959, *C. van Someren* s.n.!; Teita District: Taita Hills, *Faden* 71/91
TANZANIA. Kilosa District: Ukaguru Mts, Matandu Mt, 3–5 km WNW of Mandege Forest Station, 5 June 1978, *Thulin & Mhoro* 2958!; Morogoro District: Nguru ya Ndege Mt, 18 Sept. 1971, *Pócs & Mwanjabe* 6457/B!; Rungwe District: Ngozi crater, south side, 5 Apr. 1970, *Wingfield* 752!
DISTR. **K** 4, 7 (fide Johns); **T** 3 (fide Schippers), 6, 7; Cameroon, Bioko, São Tomé, Malawi, Mozambique, Zimbabwe, South Africa; Comoro Is., Madagascar, Mascarenes and tropical America (Peter, F.D.-O.A. records it from New Guinea and Polynesia, presumably in error)
HAB. Intermediate and montane evergreen forest, *Brachystegia* woodland influenced by mist; 950–2100 m

SYN. *Lycopodium verticillatum* L.f., Suppl. Pl.: 448 (1782): Hieron. in V.E. 2: 72, fig. 70 (1908); Sims, Ferns S. Afr. ed. 2: 325, t. 178 (1915); F.D.-O.A.: 88 (1929); Chr. in Dansk. Bot. Arkiv 7: 189 (1932); Alston in Exell, Cat. Vasc. Pl. São Tomé: 96 (1944) & Ferns W. Trop. Afr.: 12 (1959); Tardieu, Fl. Cameroun 3: 4 (1964); Schelpe, F.Z., Pterid.: 17 (1970); Schelpe & Diniz, Fl. Moçamb., Pterid.: 12 (1979); W. Jacobsen, Ferns S. Afr.: 133, fig. 75 (1983); Schelpe & N.C. Anthony, F.S.A. Pterid.: 7, fig. 4/2 (1986); J.E. Burrows, S. Afr. Ferns: 12, t. 1/3, fig. 2/3, 3a (1990)

3. **Huperzia gnidioides** (*L. f.*) *Trevisan* in Atti Soc. Ital. Sc. Nat. 17: 247 (1875); Rothm. in F.R. 54: 61 (1944); Tardieu in Fl. Madag. 13 & 13 bis: 30, fig. 5/12–15 (1971); Pic. Serm. in B.J.B.B. 53: 184 (1983). Type: Mauritius, *Sonnerat* s.n. " per Thouin"*

Pendulous or ± erect epiphyte or sometimes on rocks (sometimes terrestrial *fide* Tardieu); stems 0.2–1 m long, 2–3 times dichotomously branched. Leaves linear-oblong to lanceolate, 9–15(–24) mm long, 1.5–3(–4) mm wide, subacute to ± rounded at apex, tightly imbricate to slightly spreading, ± herbaceous to coriaceous. Fertile areas 5–12 cm long, (2–)3–4 mm wide. Sporophylls ovate-lanceolate to broadly ovate, 2–5 mm long, 1.75–2.5 mm wide, acuminate or acute or subacute at tip in forms where strobiliform areas are not distinct, coriaceous, tightly imbricate. Sporangia reniform, flattened.

var. **gnidioides**

Usually a pendulous epiphyte or from rocks. Fertile regions straight.

UGANDA. Ankole District: Bunyaruguru, Kalinzu Forest, W of Rubuzugye, 4 km NW of saw mill, 19 Sept. 1969, *Faden et al.* 69/1146! & Bushenyi, S Kasyoha-Kitomi forest, Ngozi, June 1998, *Hafashimana* 638!; Kigezi District: Rukungiri, Kayonza, Burindi forest, Ihihozo, Aug. 1998, *Hafashimana* 758!
KENYA. S Nyeri District: Mt Kenya, by R. Karute, 8 km N of Castle Forest Station, 15 Jan. 1985, *Townsend* 2209!

* Tardieu does not actually indicate this at P; Schelpe (F.S.A. Ptérid.: 9) and Øllgaard (Biologiske Skrifter 34: 47 (1989)) both give ?P, ?iso.

DISTR. **U** 2; **K** 4; ?**T** 2 (fide Peter, F.D.-O.A.); Rwanda, Burundi, Malawi, Mozambique, Zimbabwe, South Africa; Comoro Is., Seychelles, Madagascar and Mascarenes

HAB. Moist evergreen forest with *Parinari excelsa* etc.; 1300–2100 m

SYN. *Lycopodium gnidioides* L.f., Suppl. Pl.: 448 (1782); Hieron. in P.O.A. C: 90 (1895); Sim, Ferns S. Afr. ed. 2: 326, t. 177 (1915); F.D.-O.A.: 88 (1929); Chr. in Dansk Bot. Arkiv 7: 190 (1932); Schelpe, F.Z. Pterid.: 18 (1970); Schelpe, Expl. Hydrobiol. Bassin L. Bangweolo & Luapula 8 (3) Ptérid: 11 (1973); W. Jacobsen in J. S. Afr. Bot. 44: 157–163, figs. 1, A, B (1978); Schelpe & Diniz, Fl. Moçamb., Pterid.: 13 (1979); W. Jacobsen, Ferns S. Afr.: 135, figs. 77a, 77b (1983); Schelpe & N.C. Anthony, F.S.A. Pterid.: 9, fig. 3/1 (1986); J.E. Burrows, S. Afr. Ferns: 14, t. 1/5, figs. 3/5, 5a, 5b (1990)

NOTE. The Uganda material is certainly not identical with Mauritian material having more ovate, keeled sporophylls and some foliage differences, but the variation of this species needs detailed study throughout its range. Jacobsen has discussed this in the 1978 paper cited above and divides the species into three taxa, 'form' 1, 'form' 3, and 'form' 2 which he treats as *L. gnidioides* var. *pinifolium* (Kaulf.) Pappe & Rawson. He treats form 1 and 3 together i.e. they form var. *gnidioides* (he is not using form in its strict botanical sense). The Uganda material appears to be closest to form 3. Schelpe & Anthony and Burrows have not recognised var. *pinifolium* but Tardieu recognises it as a separate species *Huperzia stricta* (Bak.) Tard. (syn. *Lycopodium strictum* Bak., *L. gnidioides* L. f. var. *strictum* (Bak.) Chr., *L. pinifolium* Kaulf.). It is an erect terrestrial plant with the fertile parts drooping and ovate sporophylls and would certainly seem to be distinct at some level.

4. **Huperzia dacrydioides** (*Baker*) *Pic.Serm.* in Webbia 23: 162 (1968) & in B.J.B.B. 53: 184 (1983); Schippers in Fern Gaz. 14: 174 (1993); Faden in U.K.W.F. ed. 2: 38 (1994). Type: South Africa, Natal, *Buchanan* 184 (K!, lecto.*, photo.)

Pendulous epiphyte or less often on rocks, very rarely (not in E Africa) terrestrial; stems 20–180 cm long, dichotomously branched once or twice; basal stem horizontal, short, up to 7–8 mm thick with dense fibrous roots. Leaves closely to loosely imbricated, erect or spreading at an angle of ± 45°, coriaceous, linear-lanceolate to lanceolate, (8–)12–18 mm long, 1.5–2(–3.2) mm wide at the base, sharply acute at the apex, ± gradually passing into sporophylls or sometimes more abruptly into a strobiliform area; fertile area 2–40(–50) cm long, itself often dichotomously branched. Sporophylls oblong-lanceolate, lanceolate or triangular 2.5–13(–18) mm long, 1.8–2 mm wide, coriaceous or thinner, abruptly widened at base or not. Sporangia round, 1.5–2 mm wide.

subsp. **dacrydioides**

Stems less stiffly branched with leaves less obviously ranked and more clearly adpressed or only slightly spreading; sporophylls mostly 6–8 mm.

UGANDA. Toro District: Ruwenzori, above Minimba Camp, 22 Jan. 1902, *J.D. Loveridge* 416! & Ruwenzori, 1905, *Dawe* 660! & Bujuku Valley (? Mobuku), Nyabitaba, Apr. 1949, *Osmaston* 3984!
KENYA. West Suk District: Cherangani Mts, near Kaibwibich, 16 Mar. 1977, *P.F.S. Evans* in *Faden* 77/811!; Ravine District, Timboroa, Burnt Forest, Feb. 1933, *Mainwaring* in *Napier* 2503 (CM 5200)!; Masai District: Narok, W of Lake Magadi, Lebetero Hills, 2 July 1972, *P.E.E. Leakey* in EA 15138!

* Baker cites no specimens (save for a variety and gives "mountains of Transvaal, Natal, Zambesiland, Cameroun, Fernando Po and St. Thomas". He also states *L. passerinoides* Kuhn non H.B.K. (a misidentification). Kuhn, Fil. Afr.: 185 (1868) refers *Mann* 384 from Fernando Po and *Mann* 1408 and 2041 from Cameroon Mt to this American species. Baker makes *Mann* 2041 the type of his *L. dacrydioides* var. *brachstachys*. Schelpe labelled *Meller* s.n. (Chiradzulu Mt), clearly part of Baker's original material as lectotype on 11.6.1959 but did not publish it. Pichi Sermolli (1968) chose *Buchanan* s.n. from Natal as lectotype (actually there is a number 184 attached to the plant). (Note this is not the Buchanan who collected in Malawi although both were John). Schelpe (in F.S.A., Pterid., 1981) chose Rehmann from Transvaal, Woodbush as lectotype but there seems to be no reason for rejecting Pichi Sermolli's earlier designation.

TANZANIA. Moshi District: S slopes of Kilimanjaro, N of Moshi, 23 Feb. 1953, *Drummond & Hemsley* 1293!; Morogoro District: Uluguru Mts, Lukwangule Forest Reserve, Mar. 1955, *Mgaza* 19!; Songea District: WSW of Songea, Matengo highlands, 19 Jan. 1936, *Zerny* 349!
DISTR. U 2; K 2–6; T 2, 3, 6–8; Sierra Leone, Guinea, Liberia, Nigeria, Cameroon, Bioko, São Tomé, Congo (Kinshasa), Burundi, Sudan, Ethiopia, Malawi, Mozambique, Zimbabwe, South Africa (Natal, Transvaal); also on Comoro Is.
HAB. Evergreen intermediate and montane forest, *Ocotea–Podocarpus* etc., riverine forest, forest patches and woodland; has been recorded growing on *Ekebergia, Ocotea* and *Podocarpus*; (900–)1550–2700 m

SYN. *Lycopodium dacrydioides* Bak., Fern Allies: 17 (1867); P.O.A. C: 90 (1895); Sim, Ferns S. Afr., ed. 2: 327, t. 176 (1915); F.D.-O.A. 1: 90 (1929); Schelpe, F.Z. Pterid.: 18 (1970); Schelpe & Diniz, Fl. Moçamb., Pterid.: 12 (1979); W. Jacobsen, Ferns S. Afr.: 134, fig. 76 (1983); Schelpe & N.C. Anthony, F.S.A., Pterid.: 7, fig. 3/3 (1986); Burrows, S. Afr. Ferns: 14, t. 1/4, fig. 2/4, 4a (1990)
   L. dacrydioides Bak. var. *brachstachys* Bak., Fern Allies: 18 (1889). Type: Cameroon, Cameroon Mt, *Mann* 2041 (K!, holo.)
   L. mildbraedii Herter in Hedwigia 49: 90 (1909); Alston in Exell, Vasc. Pl. São Tomé: 95 (1944) & Ferns W.T.A.: 11 (1959); Tardieu, Fl. Cameroun 3: 11 (1964). Type: Cameroon, Cameroon Mt, Mann's Spring, *Mildbraed* 3449 (B, holo., BM, photo.)
   Urostachys mildbraedii (Herter) Nessel, Bärlappgewächse: 188 (1939) (cites Stolz 863 from T 7)
   Lycopodium brachystachys (Bak.) Alston in Bol. Soc. Brot. Sér. 2, 30: 19 (1956) & Ferns W.T.A.: 12 (1959)
   Huperzia brachystachys (Bak.) Pic.Serm. in Webbia 23: 162 (1968)
   H. mildbraedii (Herter) Pic.Serm. in Webbia 23: 163 (1968) & in B.J.B.B. 53: 184 (1983); Schippers in Fern Gaz. 14: 174 (1993)

   subsp. **dura** (*Pic.Serm.*) *Verdc.* **comb. nov.** Type: Kenya, Teita District, S of Maungu Station on Nairobi–Mombasa road, Maungu Hills, *Faden et al.* 70/180* (Herb. Pic. Serm. 25880, holo., EA, K!, iso.)

   Typically a much stiffer plant with shorter stems, more coriaceous leaves very clearly 6–8-ranked and spreading at about 45° and sporophylls very short, 2.5–4.5 mm long but material from other nearby localities merges with subsp. *dacrydioides* until the spreading leaves are the only differential character.

KENYA. Teita District: Mt Kasigau, July 1937, *Dale* in F.D. 3788! & Mbololo Hill, Mraru ridge, 17 Oct. 1970, *Faden & Githui* 70/740! & Bura bluff, Chawia Forest, 29 Aug. 1970, *Faden et al.* 70/478!
TANZANIA. Lushoto District: E Usambaras, Kwamkoro Forest Reserve, SE of Kwamkoro Tea Estate, 28 Oct. 1986, *Borhidi et al.* 86248! & Lutindi Forest Reserve, Nilo peak area, 11 May 1987, *Iversen et al.* 87470! & Amani, June 1903, *collector unknown*, A.H. 428!
DISTR. K 4/6, 7; T 3; not known elsewhere
HAB. Dry evergreen mist forest, intermediate wetter evergreen forest and *Ocotea-Cephalosphaera-Uvariodendron* etc. rain forest; 900–1550(–?1850) m

SYN. *Huperzia dura* Pic.Serm. in Webbia 27: 392, fig. 1 (1973); Schippers in Fern Gaz. 14: 175 (1993)
   H. sp. aff. *dacrydioides*; Schippers in Fern Gaz. 14: 175 (1993)

NOTE. The type is a very distinctive plant but apparently no other material has been collected in the type locality itself, so no assessment can be made on its variability in this locality. All the other material comes either from Kenya localities only 20–60 km away or from the East Usambaras, the nearest high ground S of Kasigau and only 100–150 km away. In the East Usambaras, however, apart from specimens which are clearly close to the type of *dura* (e.g. A.H. 428 cited above) there are others (e.g. *Peter* 3746, forest near Amani, Dec. 1914) from the same locality which have more adpressed leaves. I suspect that much of the variation is due to the distinctly differing climates of the localities.

* In the original description this is wrongly given as 70/108.

GENERAL NOTE. This is a very difficult species and I have been unable to sort it out satisfactorily; therefore I have taken a very wide view of it. Alston, Pichi-Sermolli, Tardieu and others have accepted that West African material belongs to two different species closely related to *H. dacrydioides* but keys to them do not work. I had thought that a subspecies based on West African material, distributed throughout most of tropical Africa, could be separated from typical material from further south based on sporophylls being lanceolate and not abruptly widened at base or shorter or even ± triangular and abruptly widened at base, but this leads to difficulties. There is no doubt that many specimens do not fit well with the wide species view e.g. *Scott Elliot* 8012 (Ruwenzori, 2700 m) and several sheets from Cameroon with very small sporophylls (e.g. *Tekwe* 236 from Buea, Mapanja) which do not agree with the taxa accepted by Alston. The single specimen seen from the Comoro Is. has short sporophylls about 4 mm long, and is very similar to material from Bioko and would undoubtedly be referred to *H. brachystachys* or *H. mildbraedii* by some authors.

5. **Huperzia holstii** (*Hieron.*) *Pic.Serm.* in Webbia 23: 163 (1968); Schippers in Fern Gaz. 14: 175 (1993). Type: Tanzania, W Usambaras, near Mashewa, Bumba, *Holst* 8814 (B, holo., BM!, fragment, K!, iso.)

Pendulous epiphyte or on rock faces; branches 12–150 cm long, 2–3.5 cm wide (including leaves) the axis fleshy and rigid, easily breaking, dichotomously branching; basal stem horizontal, creeping, up to 4 cm long. Leaves 8–10-ranked, very tightly packed, all conforming i.e. leaves and sporophylls almost identical, the fertile area not at all strobiliform, linear-lanceolate (± subulate) up to 2.25 cm long, 1.5–3 mm wide at the base, very acute at apex, spreading at up to ± 45°. Sporangia in axils of upper leaves, reniform, 2.5 mm wide.

KENYA. Masai District: 12 km ESE of Bissel, Maparasha Hills, 20 Sept. 1981, *Gilbert* 6361!; Teita District: Mbololo Hill, Ndaru ridge, 5 July 1969, *Faden et al.* 69/831! & Ngangao Forest, 9 May 1985, *Faden et al.* 294!
TANZANIA. Kilosa District: Mamiwa Forest Reserve, Ikwamba Summit, 15 Aug. 1972, *Mabberley & Balehe* 1461! Morogoro District: Uluguru Mts, Lupanga Peak, NW ridge, 26 Dec. 1931, *B.D. Burtt* 3485! & Ngeta R., above Hululu (Huala) Falls, 24 Aug. 1951, *Greenway & Eggeling* 8662!
DISTR. **K** 6, 7; **T** 3, 6, 7 (fide Luke); not known elsewhere
HAB. Mist forest with *Syzygium* and *Araliaceae*, *Newtonia* and *Macaranga*, forest on very steep almost vertical slopes with *Erythrina*, *Cussonia* etc., also wet evergreen forest of *Afrocrania*, *Garcinia*, *Ocotea*, *Allanblackia* etc.; 1350–2050 m

SYN. *Lycopodium holstii* Hieron. in P.O.A. C: 90 (1895) & V.E. 2: 71, fig. 69 (1908); F.D.-O.A.: 88 (1929)

NOTE. Schippers records this species from Mt Kindoroko in the N Pare Mts, and also mentions a sp. aff. *H. holstii* from the Nguru Mts (*Pócs & Mwanjabe* 6448) which I have not seen.

6. **Huperzia ophioglossoides** (*Lam.*) *Rothm.* in F.R. 54: 62 (1944); Tardieu in Fl. Madag. 13 & 13 bis: 40, fig. 8/9–12 (1971); Schippers in Fern Gaz. 14: 175 (1993); Faden in U.K.W.F. ed. 2: 38 (1994). Type: Mauritius (Isle de France)*, *Commerson* (P-LAM, holo.) (microfiche!)

Pendulous epiphyte or sometimes on rocks; stems (10–)30–150 cm long, dichotomously branched, 1–1.5 cm wide including leaves. Leaves loosely arranged, not entirely hiding the stem, 2–3-ranked, lanceolate, 8–12(–15) mm long, 1.5–3(–4) mm wide, acute at the apex but usually not sharp, not or scarcely coriaceous, ± erect or spreading at about 45°. Fertile parts strobiliform and usually well differentiated from leaves (but sometimes sporangia in axils of leaves indistinguishable from foliage leaves), 2–30(–40) cm long, dichotomously branched, often interrupted by a group

* Although the Lamarck type is clearly labelled Isle de France there is no material from Mauritius at K and Schelpe in FZ omits Mauritius from the distribution.

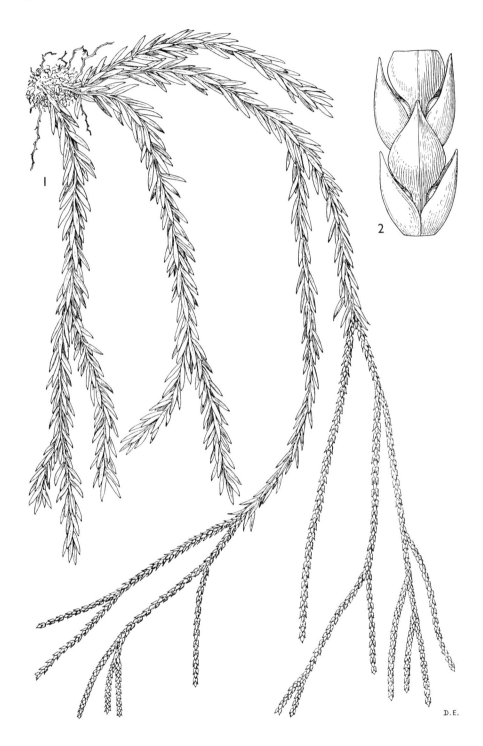

FIG. 1. *HUPERZIA OPHIOGLOSSOIDES* — **1**, habit, × ²⁄₃; **2**, sporophyll detail, × 9. 1–2 from *Brass* 16417. Drawn by Derek Erasmus, and reproduced with permission from Flora Zambesiaca.

of vegetative leaves. Sporophylls ovate to ovate-lanceolate, 2–4 mm long, 1–1.5 mm wide, ± acute, adpressed or closely imbricate but often not entirely hiding the axis, often shiny. Sporangia yellow, 1.2–1.5 mm wide. Fig. 1 (page 8).

UGANDA. Toro District: Ruwenzori, near Nyamuleju Hut, 18 June 1968, *Manum* 99! & Ruwenzori, Mubuku crossing, August 1931, *Fishlock & Hancock* 155!; Mbale District: Mt Nkokonjeru, 20 Dec. 1924, *Snowden* 951!
KENYA. Elgeyo District: Kaibwibich peak, 6 Aug. 1969, *Mabberley & McCall* 158!; South Nyeri District: Mt Kenya, Kamweti track, 26 Jan. 1969, *Faden* 69/092!; Teita District: Taita Hills, Yale Peak, 13 Sept. 1953, *Drummond & Hemsley* 4321!
TANZANIA. Moshi District: Kilimanjaro, above Kidia (Old Moshi), 18 Jan. 1997, *Hemp* 1442!; Kondoa District: Kinyassi Mt, Feb. 1928, *B.D. Burtt* 1837! & 1838!; Morogoro District: Uluguru Mts, Bondwa Hill, 23 Mar. 1953, *Drummond & Hemsley* 1769!
DISTR. **U** 2, 3; **K** 3, 4, 7; **T** 2, 3, 5–7; Bioko, Cameroon, Congo (Kinshasa), Rwanda, Burundi, Sudan, Ethiopia, Malawi, Mozambique, Zimbabwe, South Africa; Madagascar, Réunion, Mauritius (see footnote on p. 7 – sub type of this species) and Comoro Is.
HAB. Upland evergreen forest especially *Podocarpus, Schefflera, Erica*, bamboo etc., mist forest of *Newtonia, Macaranga* etc.; (1600–)1800–3250 m

SYN. *Lycopodium ophioglossoides* Lam., Encycl. Méth. Bot. 3: 646 (1789); Chr. in Dansk Bot. Arkiv 7: 192 (1932); Alston in Exell, Cat. Vasc. Pl. São Tomé: 96 (1944) & Ferns W. Tr. Afr.: 12 (1959); Tardieu, Fl. Cameroun 3: 12, t. 1/12–19 (1964); Schelpe, F.Z. Pterid.: 18, t. 3 (1970); Schelpe, & Diniz, Fl. Moçamb. Pterid.: 13 (1979); W. Jacobsen, Ferns S. Afr. 13, fig. 79 (1893); Schelpe & N.C. Anthony, F.S.A. Pterid.: 9, fig. 4/3 (1986); Burrows, S. Afr. Ferns: 16, t. 1/6, fig. 3/6, 6a (1990)
L. gnidioides sensu Peter, F.D.-O.A. 1: 88 (1929), *non* L.f.
Urostachys ellenbeckii Nessel in F.R. 36: 188, t. 175 (1934). Type: Ethiopia, 'Schoa u Galla Hochland', *Ellenbeck* s.n. (BONN-Nessel 544 pro parte, lecto.)
U. adolfi-friederici Nessel, Bärlappgewächse: 226 (1939) (*nom. inval.*) quoad *Mildbraed* 1358 (BM!, photo) (later the name was validated in Rev. Sudam. Bot. 6: 167 (1940) with *Fishlock* 83 from Ghana as type) (see note)
Huperzia ellenbeckii (Nessel) Pic.Serm. in Webbia 23: (1968)
H. afromontana Pic.Serm. in Webbia 27: 394, fig. 2 (1973) & in B.J.B.B. 53: 183 (1983); Faden in U.K.W.F. ed. 2: 38 adnot (1994). Type: Burundi, Bururi, Mt Bururi, *Lewalle* 3468 (Herb. Pic. Serm. 24932, holo.)

NOTE. Although the fertile branches are nearly always distinctly strobiliform, there are specimens where the sporangia occur in axils of leaves mostly indistinguishable from foliage leaves, e.g. *Thulin & Mhoro* 2273 (Kilosa District, Ukaguru Mts, Mamiwa Mt, 31 May 1978) and *Pócs & Mabberley* 6740/B (from same area, 30 July 1972) also *Richards* 6694 (Mbeya District, Kikondo, 21 Oct. 1956). Whether these plants would have developed the normal strobiliform areas is not clear and field studies are needed. Schippers (Fern Gaz. 14: 175 (1993)) has commented on these.
    There is at K a fragmentary specimen (Usambara, *Höfer* coll. in 1901) sent in 1923 by Nessel under the name *Lycopodium adolfi-friederici* Hert. This appears to me to be *Huperzia ophioglossoides*. Alston (Ferns W. Trop. Afr.: 12 (1959)) makes *Urostachys adophi-friederici* a synonym of *Lycopodium staudtii* (Nessel) Adams & Alston i.e. *Huperzia staudtii* (Nessel) Pic.Serm.

7. **Huperzia phlegmaria** (*L.*) *Rothm.* in F.R. 54: 62 (1944); Schippers in Fern Gaz. 14: 175 (1993). Type: Malabar and Ceylon, Dillenius, Hist. Musc. t. 61, f. 5 A, B, C (1741) (lecto.)*

Pendulous epiphyte 20–45(–75) cm long, dichotomously branched. Leaves lanceolate, flat, 9–16(–33 fide Schelpe) mm long, 2–3.5(–6 in Ceylon etc.) mm wide, widest near base, narrowly acute at the apex, narrowed or rounded at base then strongly narrowed at extreme base, somewhat coriaceous, loosely imbricate, spreading ± at right angles to stem. Fertile areas strobiliform, very sharply differentiated from

* Øllgaard (Biol. Skrifter 34: 61 (1989)) suggested this was eligible and Philcox (Rev. Fl. Ceylon Pterid., in press) has definitely chosen it. In some later reprints of Dillenius plates 61 and 62 have been changed round.

the foliage, 3–13(–24) cm long, 1.5–2 mm wide, 1–3 times dichotomously forked; sporophylls ± ovate, 1.5–2 mm long, 1.5 mm wide, only partly covering the sporangia and usually very slightly longer than them. Sporangia ± 1.3 mm wide.

Uganda. Masaka District: Sese Is., Bugala I., Kalangala, 24 Feb. 1945, *Greenway & A.S. Thomas* 7157!

Tanzania. Lushoto District: W Usambaras, Gonja, Sept. 1893, *Holst* 4279!; Morogoro District: Uluguru Mts, N slopes of Bondwa, 2 May 1970, *Pócs* 6183F!; Iringa District: Mwanihana Forest Reserve, above Sanje, 8 Sept. 1984, *D.W. Thomas* 3654!

Distr. U 4; T 3, 6, 7; Sierra Leone, Ghana, Cameroon, Gabon, Bioko, Principe, São Tomé, Malawi; Comoro Is., Mascarene Is., Madagascar (see note), also tropical Asia extending to Australia and New Zealand

Hab. *Uapaca guineensis–Schefflera* secondary forest, intermediate rain forest, riverine forest; 950–1450 m

Syn. *Lycopodium phlegmaria* L., Sp. Pl.: 1101 (1753); Hieron. in P.O.A. C: 91 (1895) & in V.E. 2: 74, fig. 72 (1901); F.D.-O.A. 1: 89 (1929); Chr. in Dansk Bot. Arkiv 7: 192 (1932); Alston in Exell, Cat. Vasc. Pl. São Tomé: 96 (1944) & in Mém. I.F.A.N.: 50: 24 (1957) & F.W.T.A. Pterid.: 12 (1959); Tardieu, Fl. Cameroun 3: 15 (1964) & Fl. Gabon 8: 9 (1964); Schelpe in F.Z. Pterid.: 20 (1970)

   Urostachys phlegmaria (L.) Herter in Bot. Arch. 3: 17 (1923)

Note. It seems unlikely *Huperzia phlegmaria* var. *tardieuae* (Herter) Tardieu in Adansonia Sér. 2, 10: 18 (1970) & Fl. Madag. 13 & 13 bis: 42, fig. 8/1–4 (1971) is distinct and Tardieu hints as much. I have therefore included Madagascar in the distribution. *Eggeling* 1576 (in F.D. 1486) from Uganda, Bunyoro District, Budongo Forest, Dec. 1954 found on *Entandrophragma utile* is sterile but has been named *Lycopodium warneckei* (i.e. *Huperzia warneckei* (Nessel) Pic.Serm.) by F. Jarrett. The differences given by Alston (in F.W.T.A. Pterid. and Mém. I.F.A.N. 50) to separate the species of the 'phlegmaria' group are slight, and certainly suspect when the immense variation accepted in India and Sri Lanka *H. phlegmaria* is considered (although several taxa may be confused there), and Dixit, Lycopodiaceae of India: 72 (1988) comments on the 'variable forms' which can be easily separated. *Eggeling* 1576 certainly agrees well with West African material named as *Lycopodium warneckei* by Alston, and Sierra Leone material has been found on the same *Entandrophragma*. Without fertile material it is not possible to confirm the identity. *H. phlegmaria* has been found on *Uapaca guineensis*.

8. **Huperzia staudtii** (*Nessel*) *Pic.Serm.* in Webbia 23: 163 (1968). Type: Cameroon, Johann-Albrechtshöhe, *Staudt* 476 (P, lecto. apparently missing, BONN-Nessel 556 pro parte, isolecto. (seen by Øllgaard))

Epiphyte said to be suberect but drooping at apex, 35–70 cm long; stems simple or dichotomously branched. Leaves lanceolate, 8–20 mm long, 2–3 mm wide, broadest near base, acute at the apex, rounded at base to point of attachment, subcoriaceous, shiny, imbricate, spreading ± at right-angles; costa ± apparent. Fertile area strobiliform, very clearly demarcated from foliage, (3–)9–23 cm long, 3–4 mm wide, unbranched or 1–3 times dichotomously branched. Sporophylls ovate, (2–)3–4 mm long, 1.5–2 mm wide, acute to a rather blunt apex, coriaceous, distinctly longer than the sporangia. Sporangia rounded reniform, ± 1.5 mm wide, flattened.

Uganda. Ankole District: Bushenyi, S Kasyoha-Kitomi Forest, Nzozi, June 1998, *Hafashimana* 637!
Distr. U 2; Ivory Coast, Ghana, Nigeria, Cameroon, Gabon, Central African Republic
Hab. Moist evergreen forest; ± 1300 m

Syn. *Urostachys staudtii* Nessel in F.R. 36: 189, t. 175 (1934)
   U. adolfi-friederici Nessel, Bärlappgewächse: 226 (1939) (anglice) & in Rev. Sudam. Bot. 6: 167, t. 13, fig. 66 (1940). Type: Ghana, near Akoase, *Fishlock* 83 (K!, holo., BONN-Nessel 553 p.p., iso.)
   Lycopodium staudtii (Nessel) Adam & Alston in Bull. Br. Mus. Bot. 1: 183 (1955); Alston in Mém. I.F.A.N. 50: 23, t. 5 (1957) & F.W.T.A. Pterid.: 12 (1959); Tardieu, Fl. Cameroun 3: 14 (1964) & Fl. Gabon 8: 9, t. 5/4–8 (1964)

2. **LYCOPODIUM**

L., Sp. Pl.: 1100 (1753) & Gen. Pl. ed. 5: 486 (1754)

Terrestrial with unequally branched subterranean creeping or climbing main stems bearing roots on underside, and with erect ascending or spreading lateral branches. Leaves uniform or not. Strobili compact, erect or pendent, simple or forked, either sessile and terminating branchlets or borne on simple or forked peduncles. Sporophylls subpeltate, peltate or chaffy scales. Sporangia reniform.

About 40 species in temperate regions and on tropical mountains; two occur in the Flora area, one known only from a single specimen.

Main stems creeping above ground, densely leafy; peduncles usually well-developed and secondary peduncles (stalks) usually (but not always) developed; leaves ending in a long hair-point, often conspicuous in tufts at end of branches and often drying orange; strobili rarely forked; sporophylls ovate, lacerate and drawn out in a long point . . . .   1. *L. clavatum*

Main stems assumed to be underground rhizomes, the erect branches very sparsely leafy at base; peduncles short and stalks absent; leaves narrowly acute but without hair-tip; strobili 2–3-forked; sporophylls lanceolate, more obviously peltate, ± entire and not drawn out into a long hair-point . . . .   2. *L. aberdaricum*

1. **Lycopodium clavatum** *L.*, Sp. Pl.: 1101 (1753); Hieron. in V.E. 2: 75 (1908); Sim, Ferns S. Afr. ed. 2: 328, t. 180 (1915); Alston in Exell, Cat. Vasc. Pl. São Tomé: 95 (1944) & in Mém. I.F.A.N. 50: 25 (1957) & Ferns W.T.A.: 12 (1959); P. Taylor, Brit. Ferns and Mosses: 40, t. 1 (bottom) (1960); Tardieu, Fl. Cameroun 3: 16 (1964); Schelpe, C.F.A.: 21 (1977); Schelpe & Diniz, Fl. Moçamb.: 14 (1979); W. Jacobsen, Ferns S. Afr.: 140, fig. 82 (1983); Pic. Serm. in B.J.B.B. 53: 185 (1983); Schelpe & N.C. Anthony, F.S.A. Pterid.: 11 (1986); J.E. Burrows, S Afr. Ferns: 18, t. 22, fig. 4/8 (1990); Schippers in Fern Gaz. 14: 174 (1993); Tsai & Shieh, Fl. Taiwan ed. 2, 1: 29, t 4/1–7 (1994); Faden in U.K.W.F., ed. 2: 38 (1994). Type: Herb. Burser XX: 49 (UPS, lecto.), chosen by Jonsell & Jarvis in Regnum Veg. 127: 63 [1993])*

Terrestrial with main stems creeping up to 1.5 m (but sometimes ± lacking), producing dichotomously branched erect stems 20–60(–80) cm tall at intervals of 5–10 cm. Leaves uniform, linear-lanceolate, 4–7 mm long, 0.5–0.75 mm wide, with a translucent hair-point 1.5–3 mm long at apex, spreading, suberect or imbricate, usually entire but some stem leaves may be ciliate; the apical hairs appear tufted at the ends of the young shoots and dry a characteristic orange-brown colour. Strobili cylindrical, 1–8 cm long, 4–6 mm wide, in groups of 2–6 (rarely solitary) at the apex of sparsely leafy branched peduncles 5–20 cm long, the stalks (secondary peduncles) (0–)0.5–5 cm long. Sporophylls broadly ovate, 3 mm long, 2 mm wide, acuminate into a long hair-point 1–2(–4) mm long, finely lacerate.

subsp. **clavatum**

Aerial erect stems from creeping main stems much branched with the branches more or less diverging; strobili on elongated simple or branched peduncles, usually not bifurcate; sporophylls shortly or long-acuminate with a hair-tip. Fig. 2 (page 12).

* Schelpe & N.C. Anthony give the type as ? Hort. Sicc. Cliff. (? BM holo.).

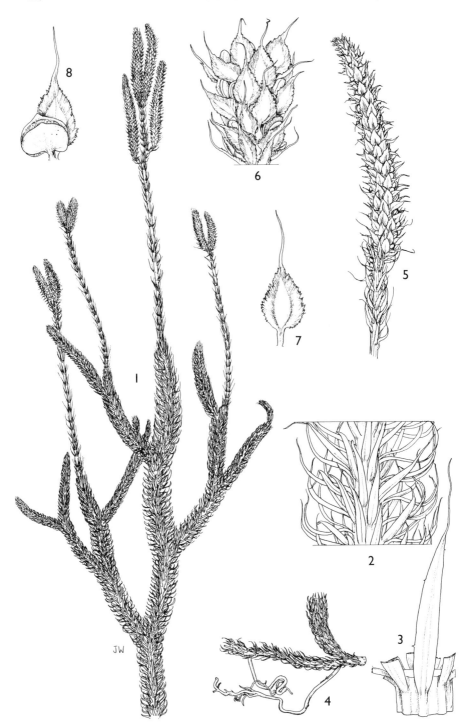

Fig. 2. *LYCOPODIUM CLAVATUM* — **1**, habit, × ²/₃; **2**, stalk/leaf detail, × 4; **3**, leaf detail, × 8;
4, root, × ²/₃; **5**, strobilus, × 2; **6**, sporophylls and sporangia detail, × 4; **7**, sporophyll, view
from outside, × 8; **8**, sporangium and sporophyll, view from inside, × 8. 1–3 from *Burtt*
2374; 4 from *Grimshaw* 93/679; 5, 7–8 from *Greenway* 3824; 6 from *Volkens* 1878. Drawn by
Juliet Williamson.

UGANDA. Toro District: E Ruwenzori, Bwamba Pass, July 1940, *Eggeling* 4028!; Ankole District: Rugongo, Nyagoma R., 9 Jan. 1971, *Rwaburindore* 505!; Mbale District: Bugishu, Elgon, Bulambuli Camp, Aug. 1934, *Synge* 984!

KENYA. Elgeyo District: Cherangani Hills, about 3 km NW of Kaibwibich, 17 July 1969, *Mabberley & McCall* 42!; Kiambu District: Kikuyu Escarpment Forest, Karamanu R. valley, along Sasamua pipeline road, 18 Aug. 1974, *Faden et al.* 74/1322!; Teita District: Taita Hills, Yale Peak, 13 Sept. 1953, *Drummond & Hemsley* 4306!

TANZANIA. Moshi District: Kilimanjaro, Bismarck Hill, 27 Feb. 1934, *Greenway* 3824!; Lushoto District: W Usambaras, Kisimba, above Mazumbai, Apr. 1916, *Peter* 16495!; Iringa District: Mufindi, Kigogo Forest, 18 Dec. 1961, *Richards* 15731!

DISTR. **U** 2, 3; **K** 3–5, 7; **T** 2, 3, 6, 7; Nigeria, Cameroon, Bioko, São Tomé, Congo (Kinshasa), Rwanda, Burundi, S Sudan, Ethiopia, Zambia, Malawi, Mozambique, Zimbabwe, South Africa; Madagascar, Mascarenes and Comoro Is.; elsewhere ± cosmopolitan

HAB. In both dry and wet habitats; bare earth, steep slopes with gritty soil, roadside banks, grassland, moist montane forest and rain forest, bamboo communities and extending into ericaceous heath e.g. *Podocarpus, Erica, Arundinaria*, rocky places, bracken scrub, swamps; in places can be dominant; (1400–)1500–3050 m

SYN. *Lycopodium clavatum* L. var. *borbonicum* Bory, Voy. Quatre Princ. Iles 2: 205 (1804); Tardieu, Fl. Madag. 13 & 13 bis: 13, fig. II/5–9 (1971); Schelpe, Expl. Hydrobiol. Bassin L. Bangweolo & Luapula 8 (3) Ptérid.: 12 (1973). Type: Réunion, Plaine des Chicots, *Bory* s.n. (P, holo.)

   Lepidotis inflexa Beauv., Prodr. aethéogam.: 109 (1805). Type: Mauritius, *Bory* s.n. (G-DEL, holo.)

   Lycopodium clavatum L. var. *inflexum* (Beauv.) Spring in Mém. Acad. Roy. Sci. Belg. 15: 90 (1842); Hieron. in P.O.A. C: 91 (1895); F.D.-O.A. 1: 89 (1929); Chr. in Dansk. Bot. Arkiv 7: 192 (1932); Schelpe in F.Z. Pterid.: 20 (1970)

   L. clavatum L. var. *inflexum* (Beauv.) Spring forma *subtilis* [sic] Chiov. in Racc. Bot. Miss. Consolata Kenia: 152 (1935). Type: Kenya, Meru District, Thingithu [Singhiso], *Balbo* 432 (TOM!, holo.)

NOTE. *Lycopodium clavatum* L. subsp. *contiguum* (Klotzsch) B.Øllg. occurs in Costa Rica, Panama, Colombia, Venezuela, Ecuador and Peru.

   R.E. & T.C.E. Fries 771 & 1894 from Mt Kenya had been named *Lycopodium trichophyllum* Desv. by the collectors but that is a Brazilian species (See C.Chr. in N.B.G.B. 9: 189 (1924)). It is now considered a synonym of *L. clavatum*.

2. **Lycopodium aberdaricum** *Chiov.* in Racc. Bot. Miss. Consolata Kenia: 152 (1935); Pic. Serm. in Webbia 27: 396 (1973); Faden in U.K.W.F.: 22 (1974); Pic. Serm. in B.J.B.B. 53: 186 (1983); Faden in U.K.W.F. ed. 2: 39 (1994). Type: Kenya, Aberdare Mts, Kinangop, *Balbo* 475 (TOM!, holo., Herb Pichi Sermolli 24103, part of holo.)

Primary stem elongate, creeping (fide Chiovenda but not preserved on type and no field notes) but owing to the nearly leafless bases of the erect shoots Faden considers it must be an underground rhizome. Erect branches 16–18 cm long, dichotomously divided 5–6 times. Leaves on lower undivided part sparse, 4–5-subverticillate, the verticils 4–10 mm apart, but dense and close on the upper branched parts. Leaves linear-lanceolate, 4–7 mm long, ± 0.7 mm wide, entire, narrowed to a fine point but without a long hair-point. Peduncle quite leafy, 2–6 cm long; secondary peduncles not developed. Strobili solitary and unbranched or conspicuously bifurcate low down with one branch again bifurcate, 1.5–7.5 cm long. Sporophylls more evidently stipitate-peltate from near the base, lanceolate, 4 mm long, 1 mm wide, narrowly acute but without a hair-point, not lacerate, virtually entire or slightly crinkly at margin.

KENYA. District uncertain: Kinangop, 20 Feb. 1910, *Balbo* 475!

DISTR. **K** 3/4; not known elsewhere

HAB. Not known but montane, possibly ± 3000 m

NOTE. Schelpe (F.Z. Pterid.: 21 (1970)) sinks this into *L. clavatum* without comment but I agree with Pichi-Sermolli and Faden that on the face of it *L. aberdaricum* is distinct but no other material has turned up from this rather well collected area. There is, I suppose always a possibility that it is some odd mutant but there are some distinctive differences. Apart from those mentioned in the key and description Chiovenda states that the spores are globose-tetrahedral similar in form and size to those of *L. clavatum* but with reticulation denser and less elevated with larger irregular areoles. Pichi-Sermolli points out that "the stalk which bears the sporangium and the blade of the leaf (sporophyll) is in the form of a fairly long subcylindric axis slightly bent downwards and broadened at the top where the blade is peltately attached; the latter is rather thick in texture and is provided basiscopically with a large greatly thickened rounded spur-like appendage".

3. **LYCOPODIELLA**

Holub in Preslia 36: 22 (1964)

Terrestrial plants with horizontal creeping or arching, sometimes branched stems rooting at short to long intervals and with erect unbranched simple (peduncles) or densely branched shoots bearing the strobili. Leaves uniform or di- or tri-morphic. Strobili distinct and compact, pendent and sessile or erect and terminating simple branches. Sporophylls subpeltate. Sporangia reniform or subglobose.

Erect shoots from creeping stem densely branched and
    Christmas tree-like bearing numerous strobili, each
    terminating an ultimate branchlet . . . . 1. *L. cernua*
Erect shoots from creeping stems unbranched peduncles bearing
    a single apical strobilus . . . . 2. *L. caroliniana*

1. **Lycopodiella cernua** (*L.*) *Pic.Serm.* in Webbia 23: 166 (1968); Tardieu, Fl. Madag. 13 & 13 bis: 8, fig. 1/9–13 (1971); Pic. Serm. in B.J.B.B. 53: 187 (1983); Schippers in Fern Gaz. 14: 174 (1993); Faden in U.K.W.F. ed. 2: 38 (1994). Type: India or E Indies; collector and locality not known, *Linnean Herb.* 1257.13 (LINN, lecto.)

Terrestrial with long creeping or looping main stems (stolons) up to 1.8 (–5 in other areas) m long, bearing erect or less often procumbent stems at intervals 0.35–1(–2) m long; roots short, up to 10 mm long. Erect stems much branched in Christmas tree-like fashion, densely leafy. Leaves subulate, 1.5–5 mm long, 0.2–0.3(–1) mm wide, spreading, curved forwards. Strobili solitary at ends of leafy branchlets, 4–15 mm long, 2 mm wide, sessile. Sporophylls pale yellowish, broadly ovate-acuminate, 1.8 mm long, 1.2 mm wide with lacerate-ciliate margins. Fig. 3 (page 15).

UGANDA. Ankole District: Rugongo, Nyagoma R., 9 Jan. 1971, *Rwaburindore* 534!; Mbale District: Bugisu, near Bukalasi, 18 Nov. 1968, *Lye* 577!; Masaka District: Sese Is., Bugala, 3 June 1932, *A.S. Thomas* 42!
KENYA. Naivasha District: S of Lake Naivasha, Hell's Gate Gorge [Njorowa], 14 Dec. 1952, *Greenway & Verdcourt* 8764!; S Nyeri District: 11 km from Embu on Embu–Sagana road, 12 Apr. 1969, *Faden et al.* 69/492!; Teita District: Mbololo Forest, 13 May 1985, *Faden et al.* 416!
TANZANIA. Lushoto District: E Usambaras, Sangerawe, 24 Oct. 1929, *Greenway* 1781!; Rufiji District: Mafia I., Ras Mbisi, 1 Oct. 1937, *Greenway* 5352!; Rungwe District: Rungwe Mt, 23 Oct. 1956, *Richards* 6753!; Zanzibar: Mazingini, 23 Feb. 1930, *Vaughan* 1260!
DISTR. **U** 2–4; **K** 3, 4, 6, 7; **T** 1–4, 6–8; **Z, P**; practically cosmopolitan
HAB. Sphagnum bogs, trackside water furrows and banks, riverine grassland, around sulphurous steam jets on lava with *Tarchonanthus, Ficus, Agarista, Erica* etc., also open parts of evergreen rain forest and steep slopes on loose gritty sand; 0–2700 m (summit of Longonot)

FIG. 3. *LYCOPODIELLA CERNUA* — **1**, habit, × ¹/₄; **2**, habit detail, × 1. **3**, strobilus, × 3; **4**, sporophyll, view from above, × 20; **5**, sporophyll from side, × 20. Reproduced from die Pflanzenwelt Afrikas 2 (1908).

Syn. *Lycopodium cernuum* L., Sp. Pl.: 1103 (1753); Hieron. in P.O.A. C: 91 (1895) & in V.E. 2: 75,
fig. 74 (1908); Sim, Ferns S. Afr., ed. 2: 327, t. 179 (1915); F.D.-O.A.: 89 (1929); Chr. in
Dansk Bot. Arkiv 7: 192 (1932); Alston in Exell, Cat. Vasc. Pl. São Tomé: 95 (1944) & in
Mém. I.F.A.N. 50: 24 (1957) & Ferns W.T.A.: 12 (1959); Tardieu, Fl. Cameroun 3: 16
(1964) & Fl. Gabon 8: 10 (1964); Schelpe, F.Z. Pterid.: 20 (1970) & Expl. Hydrobiol.
Bassin L. Bangweolo & Luapula 8 (3); Ptérid.: 12 (1973) & C.F.A. Pterid.: 19 (1977);
Schelpe & Diniz, Fl. Moçambique, Pterid.: 14 (1979); Jacobsen, Ferns S. Afr.: 138, fig. 80
(1983); Schelpe & N.C. Anthony, F.S.A., Pterid.: 11, fig. 4/1 (1986); Burrows, S. Afr.
Ferns: 18, t. 2/1, fig. 4/7 (1990)

Note. Schelpe (F.Z.) lists much synonymy not relevant to our area. Proctor (Ferns Jamaica: 29
(1985)) divides the species into var. *cernuum* and var. *curvatum* (Sw.) Hook & Grev., the latter
having denticulate ciliate leaves. *Lycopodium curvatum* Sw. had often been treated as a
separate species. Specimens from other parts of the world sometimes have hair-like cuticular
projections and a few can be found in some East African specimens. Other varieties have
been recognised but whatever the result of the future studies of the variation of this species
throughout the world which are much needed African material will remain referable to the
typical variety.

2. **Lycopodiella caroliniana** (*L.*) *Pic.Serm.* in Webbia 23: 165 (1968). Type: USA,
Carolina, Dillenius, Hist. Musc., t. 62, fig. 6 (holo.!) (the Catesby specimen on which
this figure was based is at OXF)

Terrestrial or rarely on rocks (not in East Africa) with closely creeping main stems
with adventitious roots at intervals and sometimes producing irregular pale yellowish
tubers. Leaves lanceolate to oblong or linear-lanceolate, often ± falcate, the lateral
leaves spreading horizontally, the dorsal leaves smaller or sometimes much reduced,
appressed or curving-erect, the laterals 0.5–1.5 cm long, 1–4(–?5) mm wide, the
dorsals shorter and narrower, all acuminate. Peduncles erect, borne at intervals
along the main stems and their lateral side branches, 4–35 cm long, bearing fairly
spaced narrow appressed or somewhat spreading leaves; strobilus solitary, cylindrical,
1.5–9.5 cm long (very rarely bifid or trifid at apex). Sporophylls yellowish, broadly
ovate, 3.5–6 mm long, 1.5–2 mm wide, acuminate, erose-denticulate. Sporangia
1–1.3 mm wide.

Syn. *Lycopodium carolinianum* L., Sp. Pl.: 1104 (1753)

var. **tuberosa** (*Kuhn*) *Verdc.* **comb. nov.** Type: Angola, Huila, Morro de Lopollo, *Welwitsch* 167
(?holo., BM!, K!, LISU, iso.)

Tubers present or not. Dorsal leaves similar in shape and only slightly shorter than the
laterals which are usually under 1.5 mm wide (the measurements given by Tardieu 1.5–2 mm
long for the longest leaves must be wrong).

Uganda. Masaka District: Sese Is., Bugala I., Kalangana, 2 Mar. 1933, *A.S. Thomas* 935!;
Mengo District: Kiagwe, July 1932, *Eggeling* 457 (in F.D. 793)! & W side of Kampala–Entebbe
road, 1 km N of Kisubi, 7 Sept. 1969, *Faden et al.* 69/949!
Tanzania. Bukoba District: Bukoba aerodrome, 21 June 1934, *Gillman* 62!; Rufiji District: Mafia
I., Ras Mbisi, 1 Oct. 1937, *Greenway* 5353!; Iringa District: Sao Hill, April 1959, *Watermeyer* 87!
Distr. U 4; T 1, 3, 6–8; widespread in tropical Africa and Mascarene Is.
Hab. Drier parts and edges of papyrus swamps, lakeshore and riverine grassland, forest edge
seepage areas; 5–2150 m

Syn. *Lycopodium affine* Bory, Voy. Quatre Princ. Iles 2: 204, 262 (1804); Alston in Mém. I.F.A.N.
50: 25, t. 4/10–14 (1957) & Ferns W.T.A.: 12 (1959); Tardieu, Fl. Cameroun 3: 17
(1964) & Fl. Gabon 8: 11 (1964). Type: SW Réunion, Le Volcan, Plaine des Osmandes,
Bory (P, holo.)
L. tuberosum Kuhn, Fil. Afr.: 211 (1868). Type: Angola, Huila, Morro de Lopollo, *Welwitsch*
167 (?, holo, BM, K!, LISU, iso.)

L. carolinianum L. sensu Hieron. in V.E. 2: 74, fig. 73 (1908); Sims, Ferns S. Afr. ed. 2: 329 (1915) pro parte; F.D.-O.A.: 89 (1929); Chr. Dansk Bot. Arkiv 7: 192 (1932); Ballard in Am. Fern J. 40: 83 (1950) pro parte, *non* L. sensu stricto

L. carolinianum L. var. *tuberosum* (Kuhn) Nessel, Bärlappgwächse: 274 (1939); Schelpe in F.Z. Pterid.: 21 (1970) & C.F.A. Pterid.: 23 (1977)

L. carolinianum L. var. *affine* (Bory) Schelpe in Bol. Soc. Brot. Sér. 2, 41: 214 (1967) & F.Z. Pterid.: 21 (1970) & C.F.A. Pterid.: 22 (1977); Schelpe & Diniz, Fl. Moçambique, Pterid.: 15 (1979); Jacobsen, Ferns S. Afr.: 139 (1983)

Lycopodiella affinis (Bory) Pic.Serm. in Webbia 23: 165 (1968); Tardieu, Fl. Madag. 13 & 13 bis: 9, fig. 1/5–8 (1971); Pic. Serm. in B.J.B.B. 53: 187 (1983); Schippers in Fern Gaz. 14: 174 (1993)

Lycopodium carolinianum var. *carolinianum* sensu Schelpe & N.C.Anthony, F.S.A. Pterid.: 12 (1986); Burrows, S. Afr. Ferns: 20, t. 2/4, fig. 10a (1990), *non* L. sensu stricto

NOTE. Øllgaard (in Fl. Mesoamericana 1: 19 (1995)) states that *Lycopodiella caroliniana* var. *meridionalis* (Underw. & F Lloyd) B.Øllg. & P.G.Windisch (widespread in Mexico, Central America, northern S America and West Indies) is intimately related with *Lycopodium carolinianum* var. *affine* (Bory) Schelpe and the differences between the two are uncertain. I have therefore not followed authors who have retained *Lycopodium affine* as a separate species. *Lycopodiella caroliniana* (L.) Pic.Serm. var. *caroliniana* occurs in the South eastern United States and does differ slightly in the leaves of the main stems and the peduncles. *Lycopodiella caroliniana* (L.) Pic. Serm. var. *grandifolia* (Spring) Verdc.* occurs in South Africa and Angola. Kornás (Distr. Ecol. Pterid. Zambia: 26 (1979)) suggested that a variety based on tuber production was not worth recognising and Jacobsen (Ferns. S. Afr.: 139 (1983)) sank var. *tuberosum* under var. *affine* but the use of the former at varietal rank predates the use of the other by 28 years. Badre has annotated Mauritian material as "*Lycopodiella caroliniana* var. *meridionale* (Underwood & Lloyd) Nessel & Hoehne" (sic) (the varietal gender ending and authorities refer to the taxon if placed in *Lycopodium*). But he may well be right if all the tropical material is accepted to be one taxon since the epithet *meriodionalis* used at varietal rank dates to 1927. I hesitate to sink Old World and New World taxa without a more thorough study. The species has been placed in *Pseudolycopodiella* Holub by some authors.

DOUBTFUL RECORD

Johns (Pterid. Trop. E. Afr.: 106 (1991)) records *Huperzia suberecta* (Lowe) Tardieu ('*suberectus*') from 2250 m in **U** 2, *Lors* 382. This is a species of the Azores and Madeira but Tardieu (Fl. Madag. 13 & 13 bis: 18 (1971)) records it from Réunion & Madagascar. I have not seen the specimen, nor have I ever heard of the collector.

---

* *Lycopodium carolinianum* L. var. *grandifolium* Spring in Mém. Acad. Roy. Brux. 24: 46 (1849).

INDEX TO LYCOPODIACEAE

New names validated in this volume

Huperzia dacrydiodes (*Bak.*) *Pic.Serm.* subsp. **dura** (*Pic.Serm.*) *Verdc.*
Lycopodiella caroliniana *L.* var. **grandifolia** (*Spring*) *Verdc.*
Lycopodiella caroliniana *L.* var. **tuberosa** (*Kuhn*) *Verdc.*

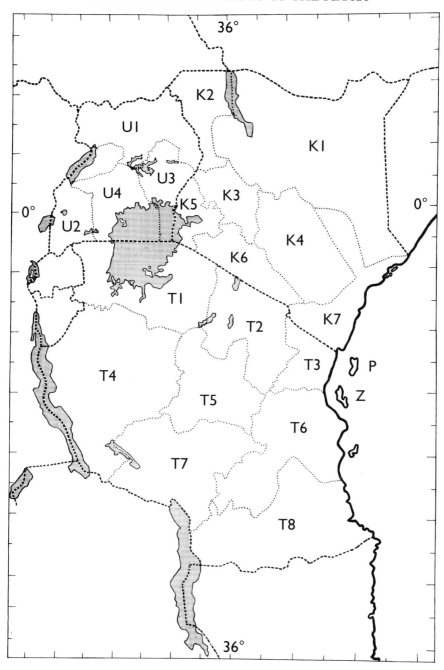

PLANTS PEOPLE
POSSIBILITIES

First published in 2005 by
Royal Botanic Gardens, Kew
Richmond, Surrey, TW9 3AB, UK
www.kew.org

ISBN 1 84246 087 0

Design by Media Resources, typesetting and page layout by Margaret Newman,
Information Services Department,
Royal Botanic Gardens, Kew.

Printed by Cromwell Press Ltd.

For information or to purchase all Kew titles please visit
www.kewbooks.com or email publishing@kew.org